Deepak Dwivedi
Vinod Singh

Laboratory Manual On Immunology and Molecular Biology

AF141889

Deepak Dwivedi
Vinod Singh

Laboratory Manual On Immunology and Molecular Biology

Step by Step Experimental Protocols, Concise and Easy to Follow

LAP LAMBERT Academic Publishing

Impressum / Imprint

Bibliografische Information der Deutschen Nationalbibliothek: Die Deutsche Nationalbibliothek verzeichnet diese Publikation in der Deutschen Nationalbibliografie; detaillierte bibliografische Daten sind im Internet über http://dnb.d-nb.de abrufbar.
Alle in diesem Buch genannten Marken und Produktnamen unterliegen warenzeichen-, marken- oder patentrechtlichem Schutz bzw. sind Warenzeichen oder eingetragene Warenzeichen der jeweiligen Inhaber. Die Wiedergabe von Marken, Produktnamen, Gebrauchsnamen, Handelsnamen, Warenbezeichnungen u.s.w. in diesem Werk berechtigt auch ohne besondere Kennzeichnung nicht zu der Annahme, dass solche Namen im Sinne der Warenzeichen- und Markenschutzgesetzgebung als frei zu betrachten wären und daher von jedermann benutzt werden dürften.

Bibliographic information published by the Deutsche Nationalbibliothek: The Deutsche Nationalbibliothek lists this publication in the Deutsche Nationalbibliografie; detailed bibliographic data are available in the Internet at http://dnb.d-nb.de.
Any brand names and product names mentioned in this book are subject to trademark, brand or patent protection and are trademarks or registered trademarks of their respective holders. The use of brand names, product names, common names, trade names, product descriptions etc. even without a particular marking in this works is in no way to be construed to mean that such names may be regarded as unrestricted in respect of trademark and brand protection legislation and could thus be used by anyone.

Coverbild / Cover image: www.ingimage.com

Verlag / Publisher:
LAP LAMBERT Academic Publishing
ist ein Imprint der / is a trademark of
AV Akademikerverlag GmbH & Co. KG
Heinrich-Böcking-Str. 6-8, 66121 Saarbrücken, Deutschland / Germany
Email: info@lap-publishing.com

Herstellung: siehe letzte Seite /
Printed at: see last page
ISBN: 978-3-659-45580-3

CONTENT

EXPERIMENT-1

Aim: Study of Antigen & Antibody Reactions.

Theory: The healthy persons are those who do not have any disease or infection. It means their body is resistant to the disease causing organisms or persons acquire immunity. Generally non-specific immunity is native or natural or hereditary. On the other hand, specific immunity is cell-mediated and humoral that is acquired by the host in response of a single or particular foreign substance, usually protein called antigen. The antigens have the ability to stimulate within the host due to the formation of homologous substance called antibodies. The antibodies are specific in their function and bind to specific antigens and thereby inactivate or kill them. The serological reactions are agglutination, precipitin formation and complement fixation.

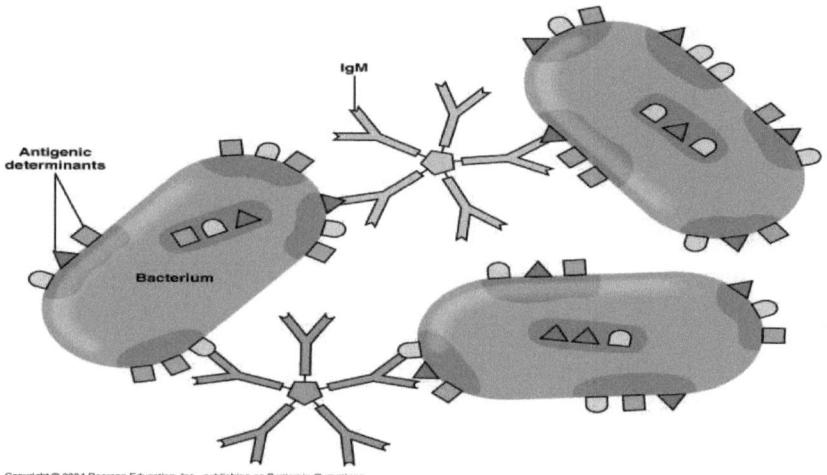

Copyright © 2004 Pearson Education, Inc., publishing as Benjamin Cummings.

A.   Slide Agglutination

Antibodies have extreme specificity for antigens, when a particular antigen is exposed to its specific antibody, it will fit to it like a hand in a glove. The ability to visualize this interaction proves a powerful tool for detecting, identifying and quantifying antibodies or for that antigen. The method of determining the presence of specific antigens is called serotyping (serological typing). The method consists of mixing of a suspension to antiserum which contains antibodies that are specific for the known antigens, formation of visible aggregate (agglutination). Similar strategy works for identification of the unknown bacterial culture.

Requirement:

Antiserum-Bacterial cultures

Antigen-for the culture

Serological tubes

Sterile saline solution

Sterile depression slides

1ml pipette

Procedure:

1. Make a phenolized saline suspension of each culture in separate serological tubes by transferring one loopful of organisms in 1ml of phenolized saline.

2. Make homogenous suspensions of microorganisms by mixing them well.

3. Take a clean depression slide and label the 3 depressions as A, B, C.

4. Transfer one loopful of each phenolized saline culture suspension to the two depressions labeled as A and B.

5. Add one drop of phenolized saline to depression C.

6. Add one drop of the antigen to the three depressions (A, B, C).

Observation:

Observe all the three depressions for positive or negative agglutination reaction within 2-3 minutes of mixing the contents.

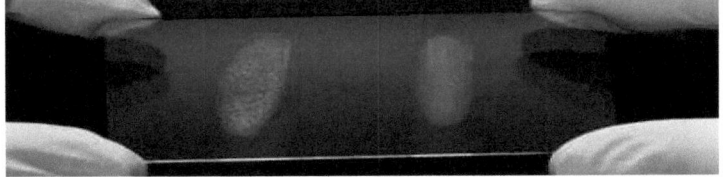

www.ssi.dk

Precautions:

1. The mixture of the bacterium and the saline should be very turbid.
2. Always use separate glass rods for mixing the contents of each depression.
3. Take care not to mix any contaminate material into the depressions.
4. Discard all slides and serological tubes in a disinfectant.
5. Slides must be examined for agglutination reaction within one or two minutes of incubation to get the true results.

B. Tube Agglutination

Grunbaum and Widal, in 1896, devised a tube test called Widal test for diagnosing typhoid fever caused by *Salmonella typhi*, a gram negative bacterium. The test is an agglutination method in which typhoid fever antibodies are detected by mixing the patients' serum with *Salmonella* cells (antigen).

Requirements:

Sample of blood serum (1:10 dilution)

Antigen-*Salmonella typhi* (1:10 dilution)

Sterile saline solution

Clean serological test tubes

1ml pipettes

5ml pipettes

Waterbath

Centrifuge

Procedure:

1. Dilute the patient's serum with saline through a series of tubes.
2. Place ten clean serological test tubes in the front row of a test tube rack and label them as 1 to 10.
3. Pipette 1ml of the patient's serum into tube 1.
4. Add 0.5ml of saline to each of the remaining nine tubes (2-10) using a 5ml pipette.
5. With a 1ml pipette, transfer 0.5ml of the serum from tube 1 to 2.

6. Mix the contents of the tube 2 by carefully drawing the liquid up into the pipette and discharging it slowly back down into the tube three times.

7. Repeat this process by transferring 0.5ml from tube 2 to 3, 3 to 4, and so on till 8 to 9 and discard the 0.5ml drawn from tube 9 instead of adding it to tube 10. Thus tube 10 has only saline and no serum will be used as a negative control.

8. Transfer 0.5ml of antigen to each tube, using a fresh 5ml pipette.

9. Shake well for complete mixing of the antigen and diluted serum.

10. Incubate all the tubes at 37°C in a Waterbath for 30 minutes.

11. Centrifuge the tubes at 2000 rpm for 7 minutes.

Observation:

Examine each tube for flakes (agglutination) against a black surface and compare each tube with negative test control.

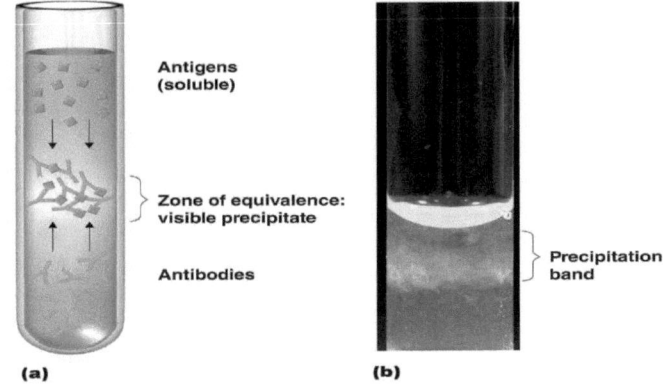

Precautions:

1. Handling of the antiserum and antigen should be done carefully.

2. Pipetting should be done precisely.

3. Take care not to mix (contaminate) any material from one depression into another.

4. Sterile glassware's should be used.

5. Discard all the serological tubes in a disinfectant.

C. **Blood Grouping**

Karl Landsteiner first of all in 1900 discovered that the red blood cells of humans were not all antigenically alike. He reported four different immunological human blood types, that is A, B, AB and O. These blood groups are called Landsteiner groups or ABO blood groups. The ABO blood groups are genetically controlled; type 'A' blood has 'A' antigens on the RBCs; type 'B' has 'B' antigens on the RBCs; type AB has both 'A' and 'B' antigens; type 'O' has neither of the antigen. ABO typing of blood is based on the principle of agglutination, type of reaction that occurs between particulate antigen and specific antibodies (agglutinins) that leads to clumping or agglutination of cells. When the cells are involve red blood cells, the reaction is termed as hemagglutination.

Requirements:

Blood typing sera: anti-A, anti-B

Disposable needle

70% alcohol

Clean glass slides

Cotton

Spirit lamp / Bunsen burner

Glass rods

Procedure:

1. Wipe the middle finger tip with cotton moistened with 70% alcohol and allow it to dry.

2. Prick the disinfected area of the finger with a sterile disposable needle.

3. Squeeze the finger and allow a drop or two of blood to fall on right and left side of a sterile slide.

4. Place a drop of antiserum-A (anti-A) on right side of the slide and antiserum-B (anti-B) on the left side.

5. Using separate glass rods mix the sera and blood drops.

Observation:

Examine the spots for agglutination for 2-3 minutes and record the results.

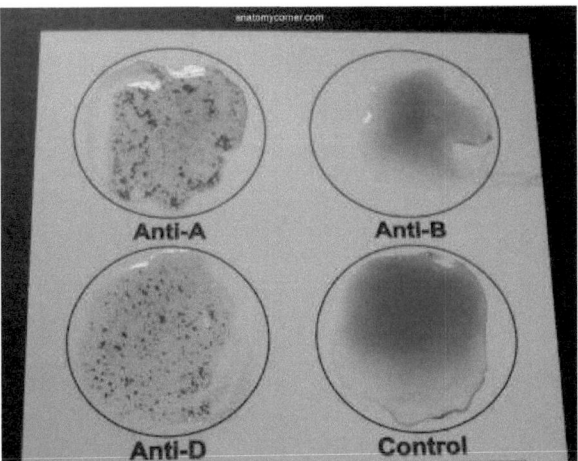

biologycorner.com

Precautions:

1. Only disposable needle should be used.

2. Sterile glass slides and rods should be used.

3. Separate glass rods should be used for mixing.

4. Mixing of the two spots with each other should be avoided.

EXPERIMENT-2

Aim : Antigen & Antibody Reaction Analysis by Enzyme Linked Immuno Sorbent Assay (ELISA)

Theory: The ELISA is an immunological technique which is used to measure either antigen or antibodies present in serum. It is also known as enzyme immunoassay (EIA). This test contains an enzyme-antibody complex that can be used as a color tracer for antigen-antibody reactions. The enzymes used most often are Horse radish peroxidase and Alkaline phosphatase, both of which release a dye (chromogen) when exposed to their substrate.

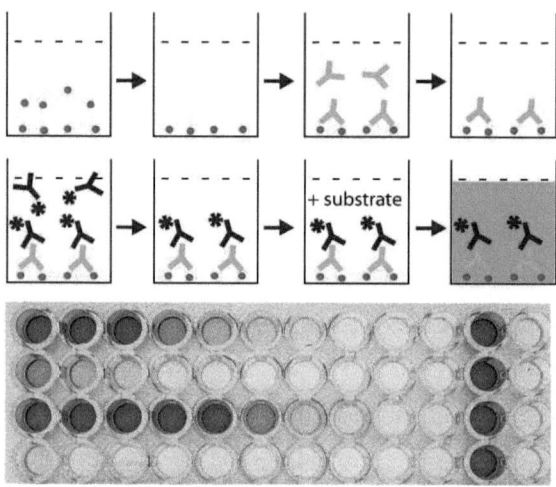

kenyaforensics.blogspot

Types of ELISA:
A) Direct ELISA
B) Indirect ELISA

A.   **Direct ELISA**

Antigen can be detected or measured by this method. Antibody (antiserum) is immobilized on the surface of wells of microtiter plate. A test antigen is added to each well and allowed to react with the bound antibody. An enzyme labeled antibody is also added which reacts with the antigen-antibody complex already present in the well and results in the development of a 'sandwich'. A chromogenic substrate is added which reacts with the enzyme and develop color.

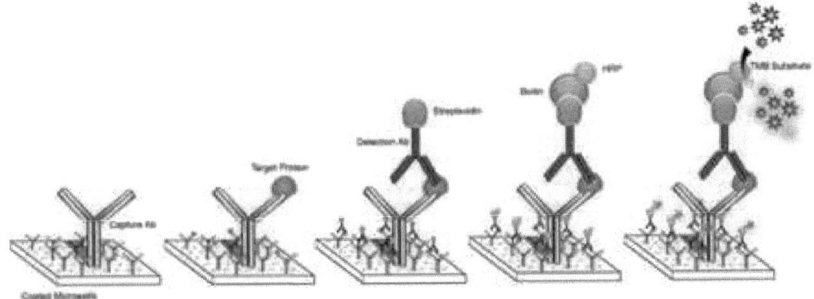

elisa-antibody.com

Requirement:

> Test antigen
>
> Antibody
>
> Enzyme linked antibody (specific for antigen)
>
> Microtiter plate
>
> Diluent (Phosphate buffer saline-PBS)
>
> Substrate (specific for enzyme)

Procedure:

1.   Dilute the antiserum to 1:500 in PBS at pH 7.2 is adsorbed to the surface of the wells on the micro titer plate.

2.   Test antigen is added to each well.

3.   Incubate the plates for 24 hours.

4.   Wash the plates three times with PBS. Unbound antigen will be washed away.

5.   Add second antibody specific for the antigens labeled with an enzyme.

6. Wash the plates three times with PBS. Unbound enzyme linked antibody will be washed away.

7. Incubate the plates for 2 hours at room temperature.

8. Add freshly prepared substrate specific for the enzyme.

9. Incubate for 30 minutes at room temperature.

Observation:

The change in color is measured visually or by measuring the optical density at 492nm using an ELISA reader to detect the desired antigen.

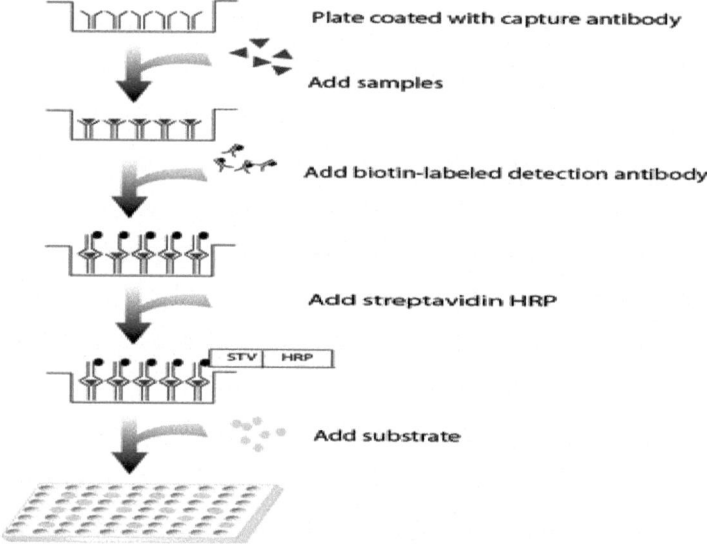

Plate coated with capture antibody

Add samples

Add biotin-labeled detection antibody

Add streptavidin HRP

STV HRP

Add substrate

myfluidics.blogspot.com

Precautions:

1. Wells of the ELISA plate should not be over filled.

2. Absorbance should be taken at the prescribed wavelength.

3. Bubbling of the antigen and antibody should be prevented while loading in the wells.

B. **Indirect ELISA**

Antibody can be detected with an indirect ELISA. It is the method of choice to detect the presence of serum antibodies against HIV. In this assay, recombinant envelope and core proteins of HIV are absorbed as solid –phase antigens to microtiter wells. Individuals infected with HIV will produce serum antibodies to epitopes on these viral proteins.

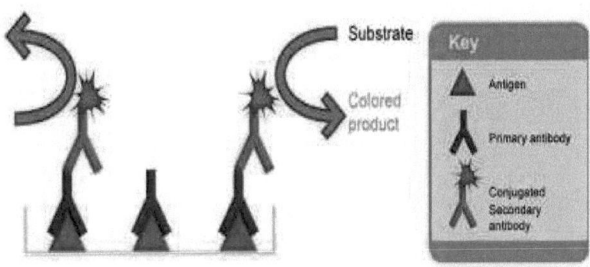

abcam.com

Requirements:

Known antigen

Primary antibody

Secondary antibody (Anti-HGG)-enzyme linked

Microtiter plate

Diluent (phosphate saline buffer-PBS)

Substrate (specific for enzyme)

Procedure:

1. Add 100µl known antigen to the wells of the microtiter plates.

2. Add primary antibody to be detected into the wells.

3. Incubate for 24 hours.

4. Wash with PBS. Unbound primary antibody will be washed away.

5. Add enzyme linked secondary antibody, Anti-HGG (an antibody that reacts with any human immunoglobulin) and allowed to react with antigen-primary antibody complex.

6. Incubate for 2 hours at room temperature.

7. Wash PBS. Unbound secondary antibody will be washed away.

8. Add specific substrate for enzymes to the wells.

9. Incubate for 30 minutes at room temperature.

Observation:

The change in color is measured visually or by measuring the optical density at 492nm using an ELISA reader to detect the desired antibody.

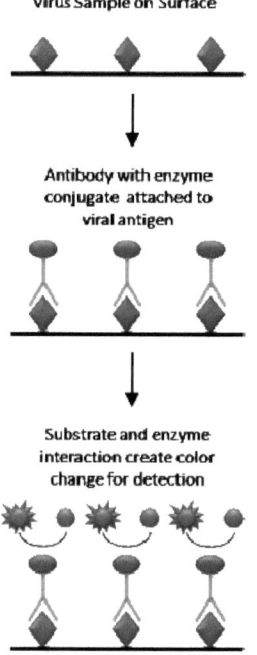

en.wikipedia.org

Precautions:

1. Wells of the ELISA plate should not be over filled.

2. Absorbance should be taken at the prescribed wavelength.

3. Bubbling of the antigen and antibody should be prevented while loading in the wells.

EXPERIMENT-3

Aim: Antigen & Antibody Reaction Analysis Radioimmunoassay (RIA)

Theory: RIA is one of the most sensitive technique for detecting antigen or antibody. It involves competitive binding of radio labeled antigen and unlabeled antigen to a high affinity antibody. The labeled antigen is mixed with antibody at a concentration that saturates the antigen binding sites of the antibody. As the concentration of unlabeled antigen increases, more labeled antigen will be displaced from the binding sites. The decrease in the amount of radio labeled antigen bound to specific antibody in the presence of the test sample is measured in order to determine the amount of antigen present in the test sample.

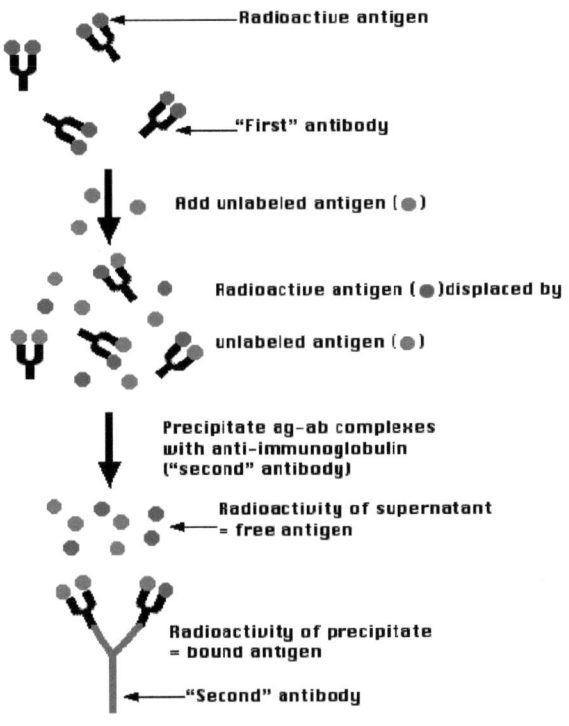

Radioactive antigen

"First" antibody

Add unlabeled antigen (●)

Radioactive antigen (●)displaced by

unlabeled antigen (●)

Precipitate ag-ab complexes
with anti-immunoglobulin
("second" antibody)

Radioactivity of supernatant
= free antigen

Radioactivity of precipitate
= bound antigen

"Second" antibody

users.rcn.com

12

Requirement:

Radioactive labeled antigen (^{125}I)

Unlabeled antigen

Sample (limited antibody)

Test tubes

Procedure:

1. Add 100μl sample antibody to the test tube.

2. Add 100μl labeled antigen to the sample antibody.

3. Further add 100μl unlabeled antigen to the labeled antigen-sample antibody complex.

4. The labeled and unlabeled Antigens compete for the limited binding sites on the Ab.

5. This competition is determined by the level of the unlabeled (test) Ag present in the reacting system. After the reaction, the Ag is separated into 'free' and 'bound' fractions and their radioactive count measured.

6. The concentration of the test Ag can be calculated from the ratio of the bound and total Ag labels, using a standard dose response curve. For any reacting system, the standard dose response or calibrating curve has to be prepared first. This is done by running the reaction with fixed amounts of Ab and labeled Ag.

Observation:

The amount of radioactivity is measured with an isotope counter or a photographic emulsion (AUTORADIOGRAPH). Two modifications of this RIA procedure have become particularly useful to the allergist. The radioimmunosorbent test (RIST) accurately measures the small amounts of IgE present in serum or body fluids. The radioallergosorbent test (RAST) is designed to measure the amount of serum IgE directed toward Antigens such as ragweed.

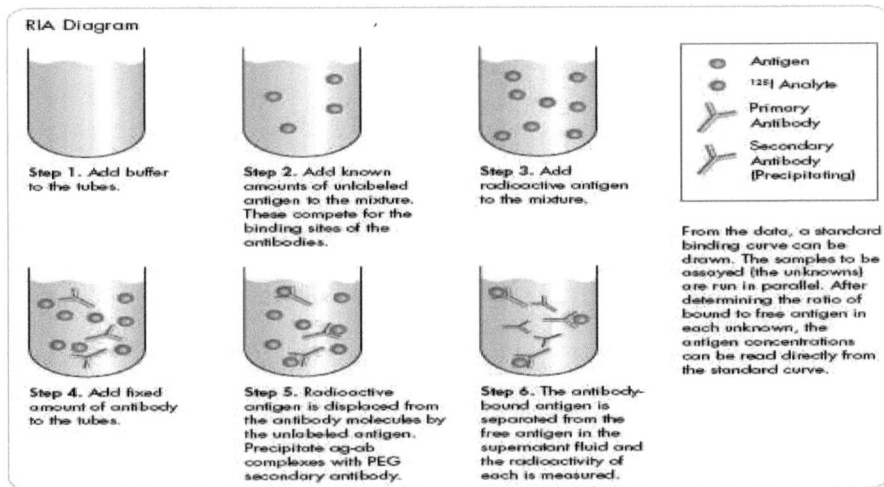

RIA Diagram

Step 1. Add buffer to the tubes.

Step 2. Add known amounts of unlabeled antigen to the mixture. These compete for the binding sites of the antibodies.

Step 3. Add radioactive antigen to the mixture.

○ Antigen
◉ ¹²⁵I Analyte
Υ Primary Antibody
Υ Secondary Antibody (Precipitating)

From the data, a standard binding curve can be drawn. The samples to be assayed (the unknowns) are run in parallel. After determining the ratio of bound to free antigen in each unknown, the antigen concentrations can be read directly from the standard curve.

Step 4. Add fixed amount of antibody to the tubes.

Step 5. Radioactive antigen is displaced from the antibody molecules by the unlabeled antigen. Precipitate ag-ab complexes with PEG secondary antibody.

Step 6. The antibody-bound antigen is separated from the free antigen in the supernatant fluid and the radioactivity of each is measured.

millipore.com Immunoassays and Multiplex

Precautions:

1. Radio labeled antigen should be used carefully.

2. Autoradiograph should be handled carefully.

EXPERIMENT- 4

Aim- Study of Antigen/Antibody reaction by Immuno Diffusion.

Immunodiffusion refers to a precipitation reaction that occurs between an antibody and antigen in an agar gel medium.

Types of immunodiffusion:

    A. Oudin technique (single diffusion)

    B. Radial immunodiffusion (rid)

    C. Ouchterlony technique

A.   OudinTechnique

Requirement:

    Test tubes

    Nutrient agar

    Antiserum

    Antigen

Procedure:

1. Take two test tubes.

2. Add single antigen and homologous antiserum into a first test tube containing melted agar at a temperature of 45°C

3. Similarly add multiple antigen and antiserum into second test tube Allow the agar containing antiserum to solidify.

4. Add an antigen containing solution to the solidified agar, spread it uniformly and refrigerate it.

5. Diffusion of Ag into the agar gel takes place during this refrigeration.

Observation:

Observe the precipitin bands formed depending on depending on the number of antigenic components in the test material, with homologous Ab. One distinct band develops for each homologous soluble Ag-Ab system present.

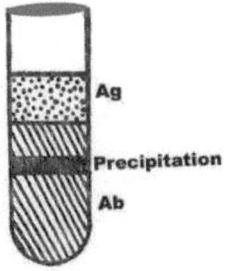

biosiva.50webs.org

B. Radial immunodiffusion (RID)

Radial immunodiffusion or Mancini method, Mancini immunodiffusion, single radial immunodiffusion assay is an immunodiffusion technique used in immunology to determine the quantity of an antigen by measuring the diameters of circles of precipitin complexes surrounding samples of the antigen that mark the boundary between the antigen and an antibody suspended in a medium, such as an agar gel. The diameters of the circles increase with time as the antigen diffuses into the medium, reacts with the antibody, and forms insoluble precipitin complexes.

Requirements:

Nutrient agar

Test antigen

Antiserum

Conical flask

Measuring cylinder

Petri dish

Gel puncher/bore core (1ml micro tip)

Procedure:

1. Prepare nutrient agar solution, add 450μl antiserum to the solution.

2. Mix by gentle swirling for uniform distribution of antiserum.

3. Pour the agar-antiserum solution onto a clean Petri dish, allow it to solidify.

4. After solidification punch the wells using a gel puncher or a 1ml micropipette tip.

5. Make more than one well and mark them separately for test antigens.

6. Make different concentrations (1:2, 1:4, 1:8, 1:16) of test antigen .

7. Load 30μl antigens in the wells as per marked.

8. Incubate for 24 to 48 hours.

9. Observe the precipitin ring formed around the well.

Observation:

As Ag diffuses into the gel, it forms a progressively widening circle of precipitation. The diameter of the precipitant ring is directly proportional to the Ag concentration in the well. This direct relationship between Ag concentration and ring diameter allows one to calculate the concentration of a known Ag.

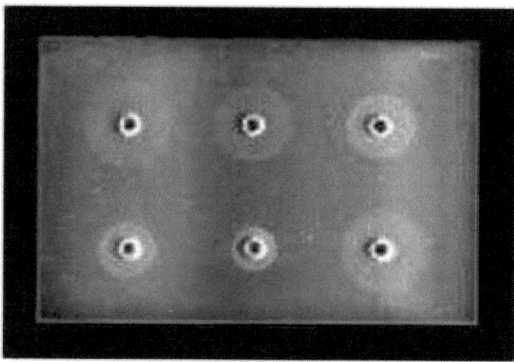

biosiva.50webs.org

Precautions:

1. Frothing of the agar solution should be avoided.

2. Enough distance should be maintained between the wells.

3. Overfilling of the wells should be avoided.

4. Agar and antiserum solution should be swirled well before pouring into the Petri dish.

5. Complete sterile condition should be maintained.

C.   Ouchterlony Technique

Ouchterlony is a gel-diffusion technique. It is an antibody-antigen precipitation test that depends on horizontal diffusion from two or more opposite sources. An agar slab is prepared and two or more wells are cut in it. One is filled with an aqueous suspension of antibody molecules, while each of the other wells is filled with a different antigen preparation. The antigen and antibody molecules diffuse toward each other and eventually interact, forming curved precipitation lines.

Requirements:

> Nutrient agar
>
> Antigen
>
> Antibody
>
> Measuring cylinder
>
> Conical flask
>
> Petri dish
>
> Micropipette
>
> Gel puncher/bore core (1ml micro tip)

Procedure:

1.   Prepare nutrient agar solution.

2.   Pour the agar solution onto a clean Petri dish, allow it to solidify.

3.   After solidification punch the wells using a gel puncher or a 1ml micropipette tip.

4.   Make more than one well and mark them separately (a center well for antigen and side wells for antibody).

5.   Load the center well with 40µl antigen and side wells with 30µl diluted antibody.

6.   Incubate for 24 to 48 hours at 37^0C.

7.   Observe the precipitin line formed around the well.

Observation:

If a V-shaped line of precipitation forms, it demonstrates that the Ab bind to the same antigenic determinants in each Ag sample and are identical. If one well is filled with a different Ag that shares some but not all determinants with the first Ag, a Y-shaped line of precipitation forms, demonstrating partial identity. In this reaction, the stem of the Y, called a spur, is formed if those Ag or Antigenic determinants absent in the first well but present in the second one react with the diffusing antibodies. If two completely unrelated antigens are added to the wells, either a single straight line of precipitation forms between the two wells, or two separate lines of precipitation form, creating an X- shaped pattern, a reaction of non-identity.

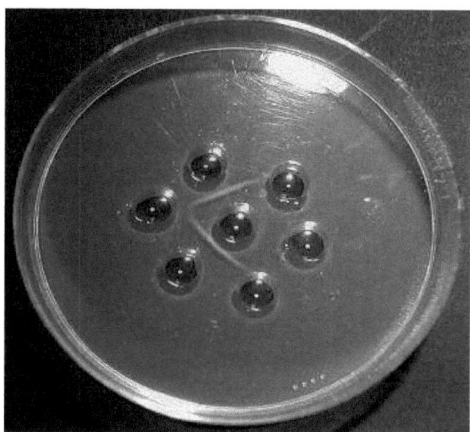

www.2ac-lyon.fr

Precautions:

1. Frothing of the agar solution should be avoided.
2. Enough distance should be maintained between the wells.
3. Overfilling of the wells should be avoided.
4. Agar and antiserum solution should be swirled well before pouring into the Petridish.
5. Complete sterile condition should be maintained.

EXPERIMENT- 5

Aim: Detection of Antigen/Antibody by Immunoelectrophoresis

Theory: When a mixture of antigens is separated by electrophoresis and detected by immunodiffusion, the technique is termed as IMMUNOELECTROPHORESIS. Immunoelectrophoresis is a general name for a number of biochemical methods for separation and characterization of proteins based on electrophoresis and reaction with antibodies. All variants of immunoelectrophoresis require immunoglobulins, also known as antibodies reacting with the proteins to be separated or characterized. Some antigen mixtures are too complex to be resolved by simple diffusion and precipitation. Greater resolution is obtained by the technique of immunoelectrophoresis in which antigen are first separated on the basis of their electrical charge, and then visualized by the precipitation reaction.

Types of immunoelectrophoresis:

A. Counter Immunoelectrophoresis

B. Rocket Immunoelectrophoresis

CIE not only depends entirely on diffusion of antigen and antibody in a gel, but also uses electrophoresis for their rapid movement. In this technique, antigen and antibody are made to move towards each other by an electric current. Visible precipitin lines develop at a point where they meet in optimal proportions. It is at least 10 times more sensitive than the double diffusion technique.

Requirements:

Antigen
Test antiserum
Glass slides
Agarose
Conical flask
Micropipettes
Micro tips
Measuring cylinder

Procedure:

1. Prepare 10ml of agarose in distilled water by heating slowly till the agarose dissolve completely.

2. Mark the end of a glass slide as +ve and –ve, so that when the glass slide is placed in the electrophoresis apparatus, the +ve mark faced towards anode and the negative mark faced towards cathode.

3. Place the glass slide on a horizontal surface. Pipette and spread 5ml of agarose onto slide, allow solidifying.

4. Cut wells with a 200µl tip. The distance between the two wells should not be more than 0.5cm.

5. Transfer the slide in the containing the gel to an electrophoresis chamber.

6. Load 30µl antigen in the negative side well and 30µl test antiserum in the positive side well.

7. Turn on the power supply, required voltage 40V.

8. After electrophoresis is complete (2-3hours), turn off the power supply.

9. Observe the precipitin line between the antigen and test antiserum

Observation:

Precipitin line indicates the presence of antibody while its absence indicates the absence of antibody.

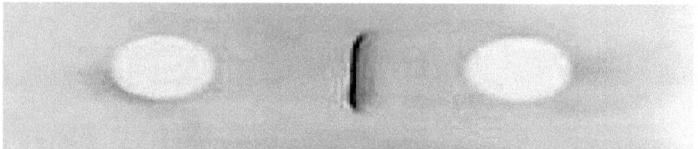

en.wikipedia.org

Precautions:

1. Antigen should always be loaded at –ve electrode and antiserum at the +ve electrode.

2. Frothing of the agarose should be avoided.

3. Proper voltage should be provided and maintained.

4. Don't disturb the slide loaded with agarose until it gets solidify.

B.   Rocket Immunoelectrophoresis

Rocket Immunoelectrophoresis is a quantitative method for serum proteins which involves electrophoresis of antigen into a gel containing antibody; the technique is restricted to detection of antigens that move to the positive pole on electrophoresis. It is a rapid way to quantify antigen in complex samples. As the antigen proteins enter the gel, they form a concentration gradient, which at some point gives the proper concentration for precipitation with the antibody in the gel. The more concentrated the antigen, the further it must run to be diluted to precipitate levels. The result is that each sample gives a "Rocket", the length of which is proportional to the concentration of antigen in the sample.

Requirements:

Agarose

Test antigen

Antiserum

Conical flask

Glass plate

Micropipette

Micro tips

Gel puncher

Electrphoretic apparatus

Procedure:

1.   Prepare 10 ml agarose in distilled water by heating slowly till the agarose dissolve completely.

2.   Add 350μl of antiserum to 10 ml molten agarose solution. Mix gently for uniform distribution of antiserum.

3.   Pour the agarose-antiserum solution onto a glass plate, allow it to solidify.

4.   Punch 4 wells with a gel puncher or a 200μl micro tip towards one end of the glass plate.

5. Place the glass plate in the electrophoresis apparatus such that the wells are towards the cathode end.

6. Load 20µl of the test antigen to the wells.

7. Connect the cord of the electrophoresis apparatus to the electric supply for the required voltage (60V).

8. Electrophoresis the samples for 2-2.5 hours, till the rockets are visible at the edges.

9. Observe the precipitation peak or rocket formed against a dark background.

Observation:

Precipitation peak or rocket can be observed against a dark background. For better visibility gel may be stained with Coomassie Brilliant Blue (CBB) for 15 to 20 minutes. Then distain with CBB for 24-32 hrs till the rocket peaks becomes visible.

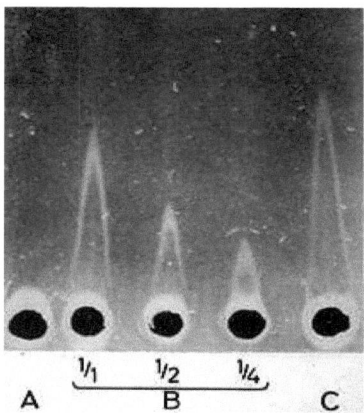

pathology.mc.duke.edu

Precautions:

1. Frothing of the agarose should be avoided.

2. Proper voltage should be provided.

3. Staining and Destaining should be done precisely.

EXPERIMENT- 6

Aim: Study of Antigen/Antibody by Immunofluorescence

Theory: Immunofluorescence is a technique used for light microscopy with a fluorescence microscope and is used primarily on biological samples. This technique uses the specificity of antibodies to their antigen to target fluorescent dyes to specific biomolecule targets within a cell, and therefore allows visualization of the distribution of the target molecule through the sample. Immunofluorescence is a widely used example of immunostaining and is a specific example of immunohistochemistry that makes use of fluorophores to visualize the location of the antibodies. Immunofluorescence can be used on tissue sections, cultured cell lines, or individual cells, and may be used to analyze the distribution of proteins, glycans, and small biological molecules.

Types of Immunofluorescence:

A. Primary/Direct Immunofluorescence

B. Secondary/Indirect Immunofluorescence

A.   Primary (direct) Immunofluorescence

Primary, or direct, immunofluorescence uses a single antibody that is chemically linked to a fluorophore. The antibody recognizes the target molecule and binds to it, and the fluorophore it carries can be detected via microscope. This technique has several advantages over the secondary (or indirect) protocol below because of the direct conjugation of the antibody to the fluorophore. This reduces the number of steps in the staining procedure, is therefore faster, and can avoid some issues with antibody cross-reactivity or non-specificity, which can lead to increased background signal.

Requirements:

Test sample

Primary antibody

2% BSA

4X PBS

2X PBS

Fluorescent stain, DAPI (4,6-diamidino phenylindole)

Fluorescent microscope

Microscopy slides

Cover slips

Procedure:

1. Add approximately 200µl of test sample on a microscopy slide.
2. Dry slightly. Do this for only 1-2 minutes.
3. Add PBS to the tissue and hydrate it until ready for next step.
4. Add 2% BSA for blocking the cells.
5. Add primary antibody diluted in 1% BSA in PBS.
6. Incubate for 45 minutes.
7. Rinse three times with 4X PBS.
8. Add DAPI, counter stain.
9. Incubate for 1 minute.
10. Rinse three times with 2X PBS.
11. Mount the slide with cover slip.
12. Examine specimen under fluorescent microscope.

Observation:

The antigen present in the test sample homologous to the primary antibody labeled with a fluorochrome combine with it and will emit fluorescence on observing under a fluorescent microscope.

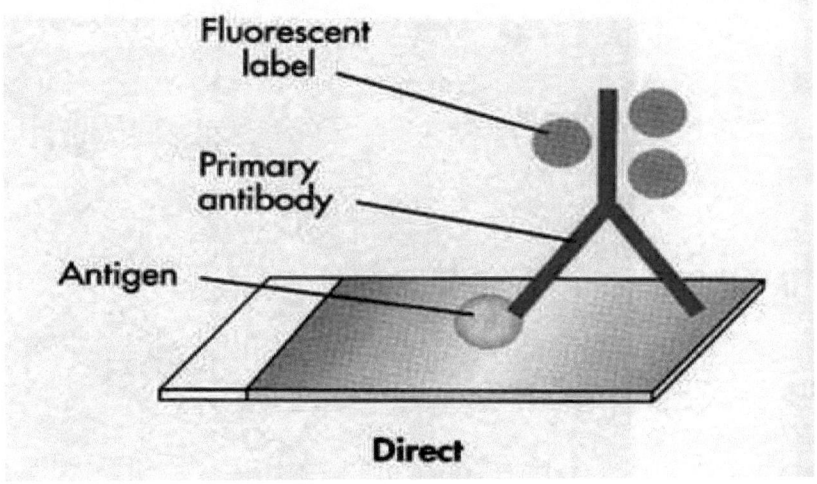

studyblue.com

Precautions:

1.   All steps should be performed at room temperature.

2.   Cells should not be dried completely.

3.   Incubation time should be maintained.

B.   Secondary (indirect immunofluorescence)

Secondary, or indirect, immunofluorescence uses two antibodies; the first (the primary antibody) recognizes the target molecule and binds to it, and the second (the secondary antibody), which carries the fluorophore, recognizes the primary antibody and binds to it. This protocol is more complex than the primary (or direct) protocol above and takes more time but allows more flexibility. This protocol is possible because an antibody consists of two parts, a variable region (which recognizes the antigen) and an invariant region (which makes up the structure of the antibody molecule). All these antibodies may therefore be recognized by a single secondary antibody. This saves the cost of modifying the primary antibodies to directly carry a fluorophore.

Requirements:

Test sample

Primary antibody

Secondary antibody

Fluorescent stain, DAPI (4,6-diamidino-2-phenylindole)

1% BSA

2% BSA

4X PBS

2X PBS

Fluorescent microscope

Microscopy slides

Cover slips

Procedure:

1. Add approximately 200μl of test sample on a microscopy slide.
2. Dry slightly. Do this for only 1-2 minutes.
3. Add PBS to the tissue and hydrate it until ready for next step.
4. Add 2% BSA for blocking the cells.
5. Incubate for 45 minutes.
6. Add primary antibody, diluted in 1% BSA, incubate for 45 minutes.
7. Rinse three times with 4X PBS.
8. Prepare fluorochrome (DAPI)-conjugated secondary antibody diluted in 1% BSA in PBS.
9. Incubate for 1 hour at room temperature in dark.
10. Rinse three times with 2X PBS.
11. Mount the slide with cover slip, observe under microscope.

Observation:

The antigen present in the test sample homologous to the primary antibody combine with it and the secondary antibody labeled with a fluorophore recognizes the primary antibody, binds to it and emit fluorescence on observing under an fluorescent microscope.

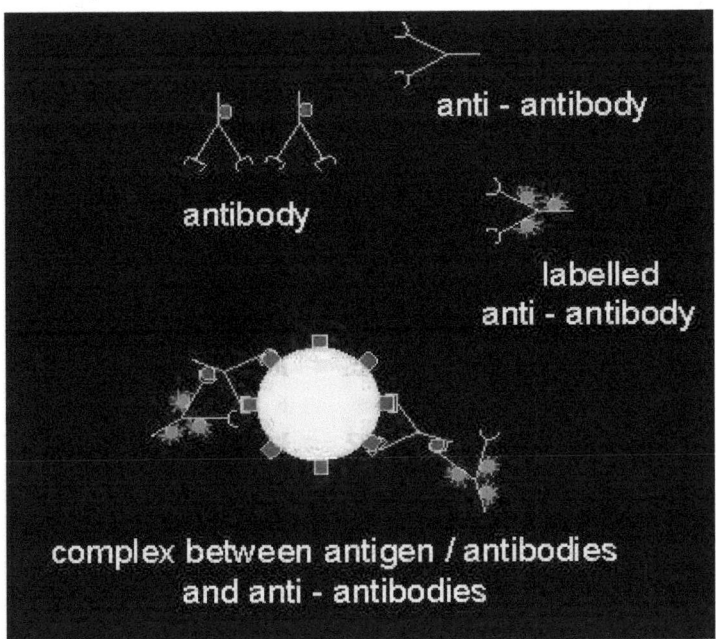

biologie.uni-hamburg.de

Precautions:

1. All steps should be performed at room temperature.

2. Cells should not be dried completely.

3. Incubation time should be maintained.

EXPERIMENT-7

Aim- Antibody Production in Rabbits

Animal Type Used   :   India White (Specific Pathogen Free)

Use: One week after arrival, having fully adjusted to the

                               facility (i.e. eating and   drinking, etc.)

Antigen Preparation:

1.   Antigen preparations include the use of adjuvants - Complete Freund's Adjuvant (CFA), Incomplete Freund's Adjuvant (IFCA), and Alum - to aid in the stimulation of the immune response. Use of either the Ribi or Titermax adjuvant systems are encouraged as a means of possibly minimizing the inflammatory response. Complete Freund's adjuvant is used in the first injection only.   The FCA and ICFA are matched in volume to the antigen, making a 1:1 mix. This mix must be thoroughly emulsified. Only ICFA can be used for booster immunizations if FCA was used for the initial immunization.

2.   All antigen preparations must be labeled with the complete name of the antigen and the number of the rabbit that is to be injected.

3.   Antigen used to inject into the rabbits is prepared by the individual labs that are using the rabbits. The following guidelines are to be used:

   a.   The antigen must be filter sterilized to reduce the amount of inflammation   at the site.

   b.   The antigen must be given vials that facilitate sterile removal of the antigen (i.e. rubber-capped serum bottles).

   c.   Purification of the protein is vital for both optimal antibody production and for the reduction of inflammation at injection sites.

   d.   The needed volume of antigen each time is 0.5 cc. This is then combined with an equal volume of adjuvant for a total of 1.0 cc.

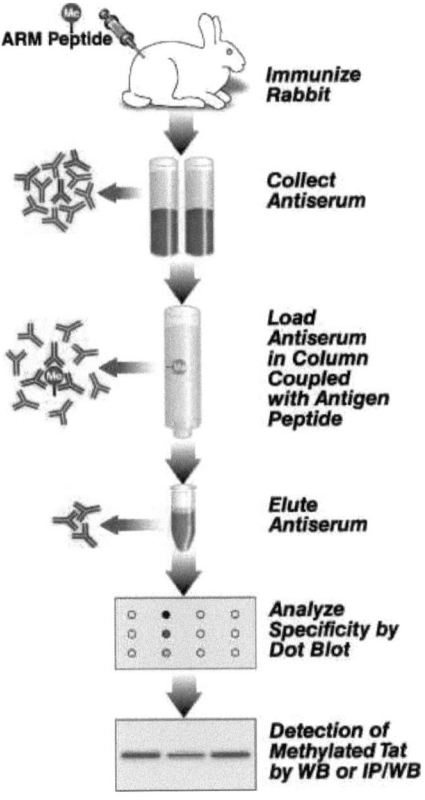

ARM Peptide

Immunize
Rabbit

Collect
Antiserum

Load
Antiserum
in Column
Coupled
with Antigen
Peptide

Elute
Antiserum

Analyze
Specificity by
Dot Blot

Detection of
Methylated Tat
by WB or IP/WB

sciencedirect.com

Injection Procedures:

1. Fractious rabbits may be tranquilized with IM acepromazine (0.1-2.0 mg/kg). Rabbits are placed in a cat-type restraint bag and transported to the procedure area.

2. Shave the area to be injected (6-8" long strip along the back extending 3" on each side of the spine).

3. Wet the area of injection with water, alcohol or disinfectant solution and wipe with a clean paper towel to remove hair and debris.

4. Injections of the antigen are given in multiple sites to stimulate the best immune response. A 22 g needle is used for injections.

5. The following sites are routinely used:

    a. SQ - multiple sites along one side of the back (0.1-0.2ml/injection) - a maximum of five sites are used, spaced an inch apart; thus a maximum volume of 1.0 cc is allowed per rabbit.

    b. IM - used for boosting with ICFA only. The maximum amount to inject into one site is 0.20ml. The biceps femoris muscle should be used.

6. The rabbits are re-immunized (boosted) at 21 day intervals until peak antibody titers are reached. ICFA is used, again at 1:1 ratio with antigen. Re-immunization injection sites should be on the opposite side of the back from the initial immunizations. Use alternate sides for additional immunizations.

7. Alternative method is to inject in popliteal lymph node only. Ideal to use when only a very small amount of antigen is available. This procedure requires following aseptic technique (sterile gloves and surgical scrub preparation of the injection site).

    a. Shave area behind the rabbit's knee. Remove excess hair with an alcohol wipe.

    b. Perform a surgical scrub using alcohol followed by iodine or chlorhexadine scrub).

    c. Apply 1% iodine solution to site prior to injection.

    d. Inject with 25g 5/8" needle, maximum amount of 0.5ml. (0.25cc antigen: 0.25cc adjuvant)

    e. The lymph node must be closely monitored for excessive inflammation. Do not inject into the same lymph node on subsequent injections.

8. The animal must be monitored closely for signs of pain, limping, anorexia, vocalization, etc. All injection sites must be closely watched for signs of inflammation and/or infection. Supportive therapy (clipping, cleaning wounds, etc.) should be done by a competent animal/veterinary technician. All lesions

must be documented in the animal's individual health record. Also, report all lesions to the facility's Attending Veterinarian by using the Notification of Animal Illness/Death Form. Analgesics (1.0 - 5.0 mg/kg butorphanol tartarate IM or SC) will be used if the lesions are deemed to cause the animal significant discomfort.

9. An animal may be euthanized at the request of the Attending Veterinarian if it meets the IACUC's Humane Endpoints in Animal Experimentation guidelines. Harvest of blood will be performed prior to euthanasia at the request of the PI.

Blood Sampling Procedures:

A. Ear Bleeds

  1. Medial Artery

    a. Administer butrophanol (10 mg/ml)/acepromazine (10 mg/ml)[1&2]. Mix at a 1:1 ratio and dose at 0.1 ml/kg BW. Inject IM.

    b. Place the rabbit in a restraint bag and shave both ears using clippers with a #40 blade.

    c. Insert a 20gx 1" short beveled needle with an appropriate size syringe (10cc usually) into the artery. Pull back on the plunger slowly until desired amount of blood is obtained. Remove needle.

    d. Apply direct pressure to the entry site until bleeding has completely stopped.

    e. Always recheck the rabbit in 10-15 minutes to assure that the ear did not start to bleed again. If the ear continues to bleed, apply digital pressure with a clean cotton ball until hemostasis is achieved.

  2. Large Volume Ear Blood Samples: A maximum of 30-40ml of blood can be obtained by this method. No more than 30-40cc of blood should be drawn from an average 8-10 lb. rabbit. A safe figure is 3.5-4cc/lb

maximum to be drawn at one time. If the maximum is drawn, that rabbit cannot be bled again for at least 3 weeks.

a.   Administer anesthetic as listed above.

b.   Place rabbit in a restraint bag.

c.   Shave both ears with clippers.

d.   Break the hub off of an 18g x 1 ½" needle. Place the needle into the median artery of the ear. Blood will pulse out of the needle when placed properly. Collect the blood in an appropriate tube - usually a 50 ml centrifuge tube.

e.   Apply direct pressure to the entry site until bleeding has completely stopped.

f.   Always recheck the rabbit in 10-15 minutes to assure than the ear did not start to bleed again.

B.   Cardiac Blood Sampling Procedure:

1.   Use ketamine (dissociative drug; 100 mg/ml) and Acepromazine (tranquilizer; 10 mg/ml) - (10:1 mixture - dose 0.35cc/10 lb. rabbit) IM. Allow approximately 5   minutes for the drug to take effect.

2.   Place the rabbit on its back on a rabbit restraint board. Secure all limbs using cotton ties.

3.   Insert an 18g x 1 ½" needle attached to 30-60ml syringe between the 4th and 5th rib on the rabbit's left side at the point where the heart beat is the strongest. Advance the needle slowly until a flash of blood is observed in the hub, then   draw back on the plunger. Alternatively, the needle may be inserted immediately to the rabbit's left of the xiphoid and directed cranially and toward the rabbit's right at a 30 to 45° angle into the heart.

4.   After filling the syringe, remove it from the needle, leaving the needle in the heart. Empty the blood into tube and reattached syringe to needle. During exsanguination the rabbit will slip deeper and deeper into an

anesthetic state   resulting in respiratory arrest followed by cardiac arrest after about 90-100cc of   blood have been drawn. This results in death. A total of 120-150cc can usually be obtained.

5.   If after exsanguination the rabbit has a heart beat but blood can no longer be obtained, the Attending Veterinarian will require euthanasia with 3-5ml of euthanasia solution[3] given by the   intracardiac   route   by trained technicians only.

6.   When the heart has stopped after exsanguination, death is ensured by creation of a bilateral pneumothorax and severing of the aorta.

Precautions:

1.   All steps should be performed at specific temperature.

2.   Animal safety guideline should be follow.

3.   Incubation time should be maintained.

EXPERIMENT-8

Aim: Isolation of Bacterial genomic DNA

Theory: DNA is obtain   from cell by most gentle method of cell rupture in the presence of EDTA to chelate $Mg2^+$ needed for DNase activity . The cell wall should be enzymatically digested by lysozyme and cell membrane using SDS. The cell disrupt should be carried out at 37°C. Protein is removed by shaking solution with phenol which denatures the proteins but not nucleic acid. Centrifuge of emulsion formed by this mixing produce in middle layer and the uppermost is aquese layer which contains the desired DNA. Finally, the deproteinised solution is mix with double volume of chilled ethanol presence at 4°C

Requirements: nutrient broth, *E. coli* culture, glass wares, eppendrof, micropipettes, tips, etc.

Instrument used: Autoclave, incubator, water bath, centrifuge, deep freezer.

Chemicals used: Nutrient broth

Tris base	1M
EDTA	0.5M
NaCl	5M
CH_3COONa	3M
SET Buffer	75mM NaCl+25mM EDTA+Tris
Lysozyme	10mg/ml
SDS	10%
Proteinase K	10mg/ml
70% C_2H_5OH	

Preparation of Stock reagents

Tric base 1 M	121.1gm/1000
EDTA 0.5M	186gm/1000
NaCl 5 M	242.5/1000
CH_3COONa 3M	246gm/1000

SET buffer	75 mM NaCl+ 25mM EDTA + 2mM Tris base
70% alcohol	70 ml C_2H_5OH+30 ml DW
10% SDS	10gm SDS+100 ml DW
Lysozyme	10 mg/ml
Proteinase K	10mg/ml

Preparation for Working reagents:-

A) TE buffer -100ml

   TE =1ml Tris+ 0.2 ml EDTA +98.8 ml DW

B) SET buffer -100ml

   2ml Tris base + 5ml EDTA + 1.5ml NaCl + 91.5ml DW

Procedure:

1. Preparation nutrient broth and inoculate bacterial culture (*E. coli*) in it

2. Inoculated the broth at 37°C for 24 hour.

3. Centrifuge the culture broth at 10000 rpm, 28°C for 10 min and collect pellet by discarding the supernatant.

4. Rinse the pellet with 1 ml TE (1 ml tris base +0.2 ml EDTA + 98.8 ml DW). Repeat this step twitch and discard the supernatant.

5. Resuspend the pellet in 1ml (500µ+ 500µl) Set buffer

6. Add 15µl lysozyme to above suspension and incubate for 30 min in water bath at 37° C.

7. Add 0.1vol (60µl) of 10% SDS and 10µl of proteinase K to above suspension and incubate in water bath at 65°C for 30 min.

8. Add 0.3vol (200µl) of 5M NaCl and equal vol. of phenol: $CHCl_3$: isoamyl alcohol-500µl (25:24:1) (Tris/water saturated phenol) incubate at room temp, for 30 min followed by gentle extraction.

9. Centrifuge at 5,000 rpm and incubate at room temp. For 15 min. remove the aqueous layer in the fresh eppendrof.

10. Add 0.1vol (100µl) 3M CH₃COONa and 1 ml vol. chilled absolute alcohol followed by gentle extraction and then incubate for 30 min at room at 4°C for 10 min , discard supernatant carefully.

11. Wash the pellet with 70% ethanol and again centrifuge at 10000rpm , 4°C for 5 min (repeat this step twice)500µl.

Precautions:

1. All reagent should be prepared in autoclave distill water.

2. The DNA is temperature sensitive, thus handled properly.

3. The entire chemical should be correctly prepared and should not be added in excess.

EXPERIMENT-9

Aim: Isolation of Fungal genomic DNA From the fresh mycelia by Phenol Extraction

Theory: In phenol extraction first solution is vigorously agitated so the protein is precipitated and the nucleic acid (RNA and DNA) is present in the aqueous layer of the solution. The result is that if the cell extract is mixed gently with solvent, and the layers then separated by centrifugation, precipitated protein molecules are left as a white coagulated mass at the interface between the aqueous and organic layers.

Requirement: Eppendrof Tube, Micropipette, Extraction buffer, iso propanol, 70%, 0.2 M Ammonium acetate, Ethanol, water bath, Tris EDTA buffer.

Solution preparation

Sewage - CHCl₃: Isoamyl alcohol) -24ml CHCl₃: 1 ml Isoamyl alcohol/70% Ethanol -70ml Ethanol + 30 ml MQ water.

Extraction buffer (w/w)

0.1 M NaCl	0.585gm
10 mM Tris Cl	0.1574gm
10 mM Na₂EDTA	0.372gm
1%SDS	1gm
MQ water / DW	100ml

Preheat at 65°C in water bath before using.

Tris EDTA: For dissolving the DNA pellet or for storage.

Tris   - 0.121 gm (0.12%)

EDTA – 0.037 gm (0.03%)

Dissolved in 100 ml distilled water

Procedure:

1. Fill a 2 ml eppendrof tube around 2/3 with ground powdered mycelium, either lyophilize or liquid nitrogen grinded (0.5-1g wet, 0.3g dry).

2. Add 0.5 ml of extraction buffer (preheated at 65°C) and incubate at 65°C water bath for 20 min.

3. Add 0.5 ml pure equilibrated phenol. Strip with toothpick and leave it for 20 minute

4. Add 0.5 of sewage (CHCl$_3$:isoamyl alcohol) 24:1 , mix well and leave it of 15 min. spin down for 10 min .at 13000rpm.

5. Remove the aqueous phase to a now tube.

6. Add the 400μlof sewage again. Vortex briefly Centrifuge at 13000rpm for 10min. remove the aqueous phase in now tube.

7. Add 0.54 volume of iso propanol. Invert gently several time. DNA ropes should be visible. Leave it at room temperature for 15 min.

8. Centrifuge as above for 10 min to pellet the DNA. Pour off the supernatant. Invert the tube for 1 min to drain.

9. Add 600μl 70% ethanol to the pellet centrifuge at 1000rpm for 5 min.

10. Remove the supernatant and dry the pellet.

11. Resupend the DNA pellet in 300μl of 0.2 M Ammonium acetate and leave it for overnight at -20°C (at least for1hour).

12. Add 600μl 100% ethanol, invert mix. Add centrifuge at 10000rpm for 15 min. dry the pellet.

13. Again wash the pellet with 70% ethanol. Spin as above.

14. Dry the pellet and resuspend in 25-50μl of TE buffer / sterile MQ water's

Precautions:

1. All reagent should be prepared in autoclave distill water.

2. The DNA is temperature sensitive, thus handled properly.

3. All the chemicals should be prepared correctly.

EXPERIMENT-10

Aim: Analysis of DNA by Agarose Gel Electrophoresis

Theory: Agarose gel electrophoresis is a method used in biochemistry and molecular biology to separate DNA or RNA molecules by size. This is achieved by moving negatively charged nucleic acid molecules through an agarose matrix with an electric field. Shorter molecules move faster and migrate further than longer ones. Agarose is usually used at concentrations between 1% and 3%. Agarose gel are formed by suspending dry agarose in aqueous buffer, then boiling and mixture until a clear solution forms then pour in the gelling tray and comb into the gel

Requirements- DNA sample, micropipettes, tips etc...

Instruments– Electrophoresis unit

Chemicals-

Tris Acetate

EDTA buffer

Agarose

EtBr solution

Bromophenol blue (dye).

Sample Preparation:

50X Tris EDTA (TE) Buffer

Tris base	242gm
Glacial acetate	57.1ml
EDTA	18.61gm
DW	1000ml
pH	8.3

6X Loading Dye

Bromophenol blue	0.25 % (W/V)
Xylene Cyanol	0.25% (W/V)
Ficol 400	15% (W/V)

Ethidium Bromide EtBr (Stock) 10mg/ml

EtBr	10mg/ml
1XTAE	1ml
Store at	20

0.4% Agarose

Agarose	0.877
1XTAE	100 ml

Procedure

Preparing 0.8% agarose gel:-Prepare 1X TAE/TBE by diluting 50X of the TAE or TBE buffers. For 0.8% gel concentration,

1. Dissolved the Agrarose by heating until the solution becomes the clear. Add 6µl Ethidium Bromide.
2. Prepare gel tray .and place comb in gel tray about q inch from one end of tray and placed the comb vertically such that the teeth and about 1-2mm above the surface of tray.
3. Pour gel solution into tray to a depth of about 5mm.
4. After solidification, remove the combs and put the gel in the electrophoresis tank.
5. Fill the tank with 1X TBE buffer.
6. Add 10 µl of 6X gel loading dye for every 5 µl of DNA solution. Mix well an1 µl DNA sample per well.
7. 7. Electrophorus   at 50v, until dye marker migrates at appropriate distance.
8. Observe the bands on gel under UV light.

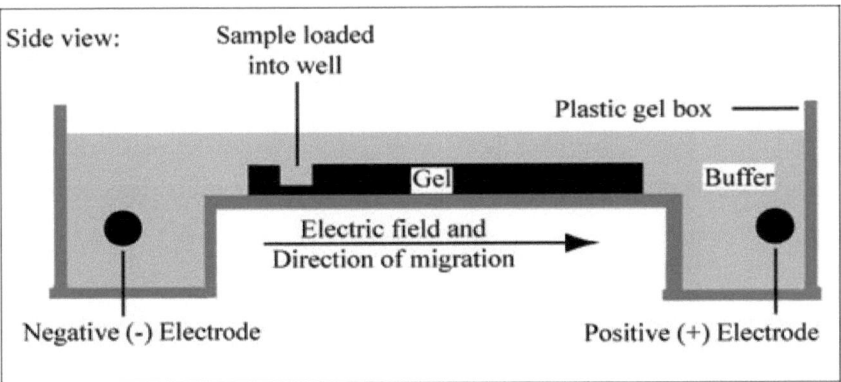

Precautions:

1. EtBr is carcinogenic Compound so gloves should always be used while handling it.

2. Weighing of reagent should be done properly so buffer should be prepared accurately.

3. Sterilization should be used for avoiding fluctuation.

4. Loading of sample should be done carefully.

EXPERIMENT-11

Aim: Isolation of Plasmid

Theory: Plasmid is a double stranded, circular extra chromosomal DNA of bacterium. It is used in recombinant DNA experiments to clone genes from other organisms and make large quantities of their DNA. Plasmid can be transferred between same species or between different species. Size of plasmids range from 1-1000 kilo base pairs. Plasmids are part of mobilomes (total of all mobile genetic elements in a genome) like transposons or prophages and are associated with conjugation. Even the largest plasmids are considerably smaller than the chromosomal DNA of the bacterium, which can contain several million base pairs. Isolation of plasmid DNA from *E. coli* is a common routine in research laboratories. The protocol often referred to as a plasmid "mini-prep," yields fairly clean DNA quickly and easily.

Requirements- Nutrient broth, E-coli Culture, glassware, eppendroff, micropipettes, tips. Etc

Instruments- Autoclave, Incubator, Water bath, Centrifuge, Deep freezer .etc

Solutions:

Solution 1:	**per 500 ml:**
50 mM glucose	9 ml 50% glucose
25 mM Tris-HCl pH 8.0	12.5 ml 1 M Tris-HCl pH 8.0
10 mM EDTA pH 8.0	10 ml 0.5 M EDTA pH 8.0
Add H$_2$O to 500 ml.	

Solution 2:	**per 500 ml:**
1% SDS	50 ml 10% SDS
0.2 N NaOH	100 ml 1 N NaOH
Add H$_2$O to 500 ml.	

Solution 3:　　　　　　　　　　　**per 500 ml:**

3 M K+　　　　　　　　　　　　　　300 ml 5 M Potassium acetate

5 M Acetate　　　　　　　　　　　　57.5 ml glacial acetic acid

Add H$_2$O to 500 ml.

Tris EDTA Buffer　　　　　　　　**per 100 ml:**

10 mM Tris-HCl pH 8.0　　　　　　　1 ml 1 M Tris-HCl pH 8.0

1 mM EDTA　　　　　　　　　　　　0.5 ml 0.5 M EDTA pH 8.0

Add H$_2$O to 100 ml.

Optional: RNAse can be added to TE at final concentration of 20 µg/ml.

Procedure:

1. Fill a microcentrifuge tube with saturated bacterial culture grown in LB broth + antibiotic. Spin tube in microcentrifuge for 1 minute, and make sure tubes are balanced in microcentrifuge. Dump supernatant and drain tube briefly on paper towel.

2. Repeat step 1 in the same tube, filling the tube again with more bacterial culture. The purpose of this step is to increase the starting volume of cells so that more plasmid DNA can be isolated per prep. Spin tube in microcentrifuge for 1 minute. Pour off supernatant and drain tube on paper towel.

3. Add 0.2 ml ice-cold Solution 1 to cell pellet and resuspend cells as much as possible using disposable transfer pipette.

   o Solution 1 contains glucose, Tris, and EDTA. Glucose is added to increase the osmotic pressure outside the cells. Tris is a buffering agent used to maintain a constant pH (= 8.0). EDTA protects the DNA from degradative enzymes (called DNAses); EDTA binds divalent cations that are necessary for DNAse activity.

4. Add 0.4 ml Solution 2, cap tubes and invert five times gently. Let tubes sit at room temperature for 5 minutes.

- o Solution 2 contains NaOH and SDS (a detergent). The alkaline mixture ruptures the cells, and the detergent breaks apart the lipid membrane and solubilizes cellular proteins. NaOH also denatures the DNA into single strands.

5. Add 0.3 ml ice-cold Solution 3, cap tubes and invert five times gently. Incubate tubes on ice for 10 minutes.

- o Solution 3 contains a mixture of acetic acid and potassium acetate. The acetic acid neutralizes the pH, allowing the DNA strands to renature. The potassium acetate also precipitates the SDS from solution, along with the cellular debris. The *E. coli* chromosomal DNA, a partially renatured tangle at this step, is also trapped in the precipitate. The plasmid DNA remains in solution.

6. Centrifuge tubes for 5 minutes. Transfer supernatant to fresh microcentrifuge tube using clean disposable transfer pipette tips. Try to avoid taking any white precipitate during the transfer. It is okay to leave a little supernatant behind to avoid accidentally taking the precipitate.

- o This fractionation step separates the plasmid DNA from the cellular debris and chromosomal DNA in the pellet.

7. Fill remainder of centrifuge tube with isopropanol. Let tube sit at room temperature for 2 minutes.

- o Isopropanol effectively precipitates nucleic acids, but is much less effective with proteins. A quick precipitation can therefore purify DNA from protein contaminants.

8. Centrifuge tubes for 5 minutes. A milky pellet should be at the bottom of the tube. Pour off supernatant without dumping out the pellet. Drain tube on paper towel.

- o This fractionation step further purifies the plasmid DNA from contaminants. This is also a good place to stop if class time is running out. Cap tubes and store in freezer until next class period.

9. Add 1 ml of ice-cold 70% ethanol. Cap tube and mix by inverting several times. Spin tubes for 1 minute. Pour off supernatant (be careful not to dump out pellet) and drain tube on paper towel.

   o Ethanol helps to remove the remaining salts and SDS from the preparation.

10. Allow tube to dry for ~5 minutes. Add 50 ul TE to tube. If needed, centrifuge tube briefly to pool TE at bottom of tube. DNA is ready for use and can be stored indefinitely in the freezer.

Precautions:

1. All reagent should be prepared in autoclave distill water.

2. The DNA is temperature sensitive, thus handled properly.

3. The entire chemical should be correctly prepared and should not be added in excess.

4. Loading of sample should be done carefully

5. Although making your own solutions is much cheaper, you can also order a kit containing all the solutions primate from Carolina Biological. Enough reagents are provided for ~ 40 mini-preps. Plasmid DNA Isolation Reagent System.

6. In the classroom, we have been able to complete the isopropanol step (step 8) by the end of the class period (50 minutes). The next day, the rest of the procedure can be completed, and the next experiment, such as a restriction enzyme digest, can be set up.

EXPERIMENT-12

Aim*:* Amplification DNA by using Thermal cycler.

Theory: Polymerase Chain Reaction (PCR), invented by Kary B. Mullis, at the Cetus Corporation, who was awarded the 1993 Nobel Prize for chemistry for PCR, is a technique to exponentially amplify in vitro a small quantity of a specific nucleotide sequence in the presence of template sequence, two oligonucleotide primers that hybridize to opposite strands and flank the region of interest in the target DNA, a thermostable (taq) DNA polymerase. The reaction is cycled involving template denaturation, primer annealing, and the extension of the annealed primers by DNA polymerase until enough copies are made for further analysis.

Polymerase Chain Reaction is a molecular biology technique which is used to amplify the number of copies of specific region of DNA enzymatically without using a living organism such as yeast, *E.coli*. This is a powerful technique that produces millions of copies of DNA from a single copy, with high accuracy, specificity and in very short period of time

PCR Technique:

The PCR involves three steps: - Denaturing, Annealing and Extension.

Denaturing:- At this step, the two strands of DNA are separated. The temperature ranges from 90ºC to 98ºC.

Annealing:- At this step, the complementary primers attach themselves to the ends of the single stranded target DNA. The temperature ranges from 40ºC to 60ºC.

Extension:- AT this step, the thermo stable Taq DNA polymerase is used to synthesize new DNA strands, by using the primers and the template DNA as the basis. Extension temperature ranges from 68ºC to 72ºC.

Requirements:

Equipment: Thermocycler, Power supply Unit

The PCR reaction components are

Template DNA: - 0.05-1.0 ng of template DNA which will be amplified by the PCR.

dNTPs: - Deoxynucleotides to provide both the energy and nucleosides for the synthesis of DNA. It is important to add equal amount of each nucleotide (dATP, dTTP, dCTP and dGTP) to the master mix to prevent mismatches of bases.

PRIMERS :- Primers are short pieces of DNA (20-30 bases) that bind to the DNA template allowing Taq DNA polymerase enzyme to initiate incorporation of the deoxynucleotides forward and reverse primers are required, which can be either specific or universal. 16s reverse and 16s Forward primer used Bacteria and ITS-1 and ITS-4 Universal primers use for Fungi.

Taq Polymerase: A heat stable enzyme that adds the polymerase deoxynucleotides to the DNA template and synthesize new strands.

Buffer: reaction buffer, which provides a suitable chemical environment for the DNA polymerase and to keep the reaction at the proper pH.

Reaction Mixture

For 50.00 µl

S.NO.	Chemical	Quantity
1	10X Taq buffer with$(NH_4)_2SO_4$buffer	5µl
2	25mM $MgCl_2$	6µl
3	2 mM dNTPs	4µl
4	16s-R /ITS-1(Primer 1)	4µl
5	16s-F/ ITS-4 (Primer 2)	4µl
6	Taq polymerase	0.6µl
7	Milli q water	21.4µl
8	DNA	5µl

Procedure:

1. Label the PCR tubes.

2. Add the PCR reagents one by one.

3. Then add the 10X reaction buffer followed by dNTPs.

4. Add the primers.

5. Then add DNA and finally Taq polymerase.

6. Spin the mix for a few seconds to mix the reagents properly.

7. Place the tube in the thermo cycle and run the following program.

PCR conditions: - Bacteria.

S.NO.	STEPS	TEMERATURE	TIME
1	Complete denaturation	92 °C	2 min.
2	Denaturation	92 °C	1 min.
3	Annealing	48 °C	30 sec.
4	Extension	72 °C	2 min.
5	Step 2-4 are repeated 30-35 Times		
6	Final extension	72 °C	6 min.
7	Hold	4 °C	forever

PCR conditions: - Fungi

S.NO.	STEPS	TEMERATURE	TIME
1	Complete denaturation	94 °C	5 min.
2	Denaturation	94 °C	30 sec.
3	Annealing	55 °C	30 sec.
4	Extension	72 °C	1 min.
5	Step 2-4 are repeated 30-35 Times		
6	Final extension	72 °C	10 min.
7	Hold	4 °C	forever

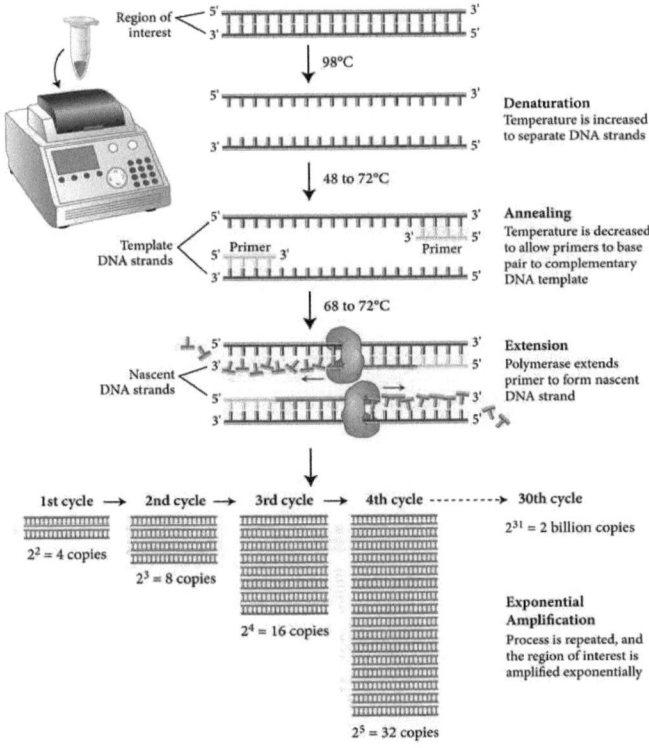

neb.com/applications/dna-amplification-and-pcr

Analysis of PCR reaction: - A PCR is considered successful if,

   a) A product is formed, and

   b) The size of the product is correct.

PCR products are visualized by agarose or polyacrylamide gel electrophoresis. The product purity can be confirmed by sequencing.

Precautions –

   1. All reagent should be prepared in autoclave distill water.

   2. The DNA is temperature sensitive, thus handled properly.

   3. All the chemical should be correctly prepared and should not be added in excess

   4. Loading of sample should be done carefully

.EXPERIMENT-13

Aim: Restriction fragments length polymorphism (RFLP).

Theory: Each individual has a unique and characteristic combination of nucleotides in their DNA. This unique combination can be analyzed and used for definitive identification. To do this, the person's DNA is first cut into small fragments by the action of restriction endonucleases. These are enzymes (derived from bacteria largely) which cut DNA into pieces based on unique sequences recognized by the different enzymes. Different pieces of DNA are produced by using combinations of different enzymes, and the pieces produced by each person's DNA are unique. The DNA fragments produced are then separated by gel electrophoresis to produce patterns. That will help in molecular determination.

Requirements: Buffer, DNA sample, Electrophoresis unit, power supply, Restriction Enzyme

Reaction Mixture for RFLP

For 15 μl

S.No.	Substances	Quantity
1	10x Buffer	1.5μl
2	Restriction enzyme	1μl
3	DNA (PCR amplified product)	8μl
4	Milli q water	4.5μl

Procedure:

1. Take the 8μl ITS /16s amplified region of DNA.
2. Transfer into PCR tube than add 1 μl Restriction Enzyme and add 1.5μl, 10X buffer and 4.5 μl MQ water.
3. Incubate the DNA with enzyme at 37°C (temperature varies according Enzyme to Enzyme) for 3 hour.

4. Mix fragmented DNA with loading dye and load on1.5X agarose gel for 4hr at 60 volt 105 amp.

5. Analyze under the Gel doc.

Precautions:

1. All reagent should be prepared in autoclave distill water.

2. The DNA is temperature sensitive, thus handled properly.

3. All the chemical should be correctly prepared and should not be added in excess.

4. Loading of sample should be done carefully

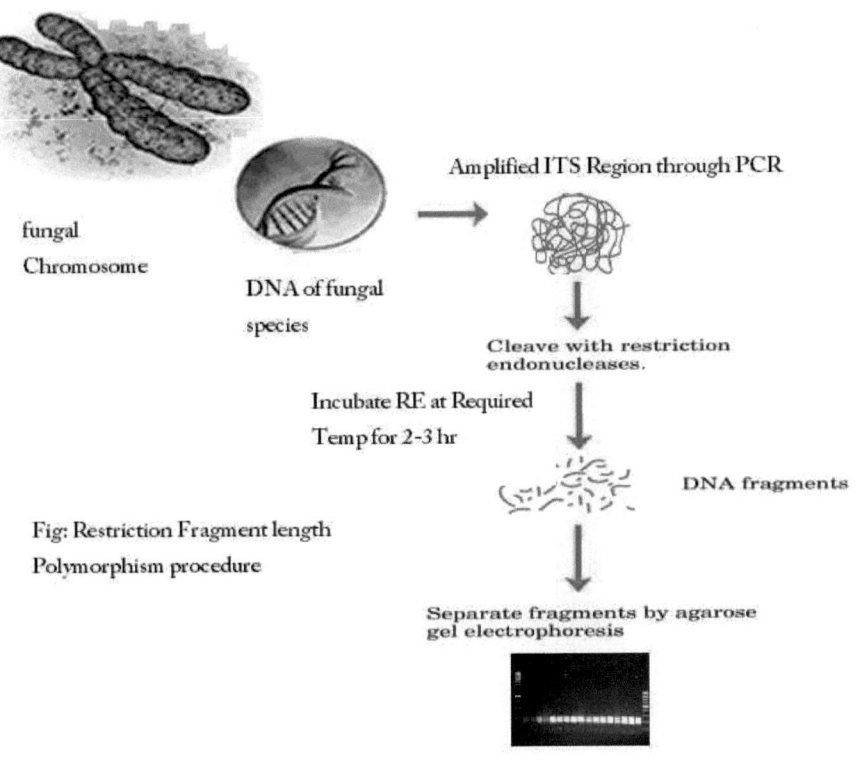

Fig: Restriction Fragment length Polymorphism procedure

Bio.com/applications/dna-rflp

EXPERIMENT-14

Aim: Random Amplified Polymorphic DNA (RAPD)

Theory: Random primers are used to amplify genomic DNA. Patterns of bands may be different for individuals in a population or closely related species. For RAPD PCR primer design. RAPD reactions are PCR reactions, but they amplify segments of DNA which are essentially unknown to the scientist (random). Often, PCR is used to amplify a known sequence of DNA. Thus, the scientists chooses the sequence he or she wants to amplify, then designs and makes primers which will anneal to sequences flanking the sequence of interest. Thus, PCR leads to the amplification of a particular segment of DNA

Requirements:

Equipment: Thermocycler supply unit

Water: Sterile de-ionized or distilled water should be used for preparing all the reagents.

Reaction buffers: Assay buffer for Taq DNA polymerase (supplied by the manufacturer of Taq DNA Polymerase).

Deoxynucleoside triphosphates (dNTP'S): 2.5 mM each of dCTP, dATP, dTTP, dGTP. Readymade solutions of dNTPS are available from many manufactures. Store at -20°C

Magnesium chloride: 25mM stock and store at -20^{0}C

Genomic DNA: 5-25 ng/ml stocks. DNA of sufficient quality can be obtained in coconut by using SDS protocol.

Procedure:

Assemble RAPD reactions as follows:

2.5µl DNA stock (25ng/µl)

2.5µl Assay buffer containing 2.5mM MgCl2 (2.5mM)   1µl MgCl2 stock (1.5mM)

1µl primer stock (25pmol)

4µl dNTPS (400µM)

1µl Taq polymerase (1U)

Sterile water to make 25 µl

Wear gloves throughout RAPD reaction preparation procedure. Assay buffer, dNTPs, MgCl2 and primer solution are thawed from frozen stock. Keep the assembled reaction in themocycler for amplification.

Amplify DNA in themocycler: Cycling conditions may be modified depending on the thermocycer used

Temperature profile

General cycling steps followed are:

PCR conditions:-

S.NO.	STEPS	TEMERATURE	TIME
1	Complete denaturation	94 °C	5 min.
2	Denaturation	94 °C	1 min.
3	Annealing	55 °C	1 min.
4	Extension	72 °C	2 min.
5	Step 2-4 are repeated 30-39 Times		
6	Final extension	72 °C	10 min.
7	Hold	4 °C	Forever

After the reaction, DNA is analyzed through gel electrophoresis.

Agarose gel electrophoresis

Reagents for agarose gel electrophoresis:

Agarose, TBE/TAE buffer, ethidium bromide, gel loading dye,

To prepare 100ml of a 0.7% agarose solution, measure 0.7g agarose into a glass beaker or flask and add 100ml 1X TBE or TAE.

Microwave or stir on a hot plate until agarose is dissolved and solution is clear.

Allow solution to cool to about 55° C before pouring. (ethidium bromide can be added at this point to concentration of 0.5µg/ml

Place the comb in gel tray.

Pour 50° C gel solution into tray to a depth of about 5mm. Allow the gel to solidify for about 20 min at room temperature.

To run, gently remove the comb, place the tray in electrophoresis chamber, and cover (just until wells are submerged) with electrophoresis buffer (the same buffer used to prepare the agarose).

To the RAPD sample from refrigerator, add 1µl of 6%gel loading dye for every 5 µl of DNA solution. Mix well. Load 20µl of DNA per well. Load also the DNA size standards along side RAPD reactions.

Connect the electrodes to the power pack. and electrophoresis at 50-150Volts until the bromophenol blue dye has reached three fourth of the gel length.

Stain the gel with ethidium bromide (if not already included in the gel).

Examine the gel under UV light (transilluminator).

Depending on the objective of the experiment make a note of polymorphism, segregating bands, and appearance of overall pattern with in fingerprint.

Bands may be sized by comparison to molecular weight standards. The standards should be used to generate a standard curve for interpolation.

After you have run the gel, obtain a photograph, and label and measure the migration of the DNA bands. Make a standard curve plot of the known size markers, and determine the size of the marker bands.

Gel Interpretation:

Bands are sized and matched directly on gels, or photographic films, or photocopies on transparency overlays.

Note the presence and absence of bands.

Analyze the data using computer software NTSYS / RAPDistance

Precautions:

1. All reagent should be prepared in autoclave distill water.

2. The DNA is temperature sensitive, thus handled properly.

3. All the chemical should be correctly prepared and should not be added in excess.

4. Loading of sample should be done carefully.

5. Ethidium bromide is a mutagen and a probable carcinogen. Wear gloves when working with ethidium bromide solutions.

6. Also use care not to contaminate the work area with the solution.

7. UV light is damaging and must be used with caution. UV light causes burns and can damage the eyes.

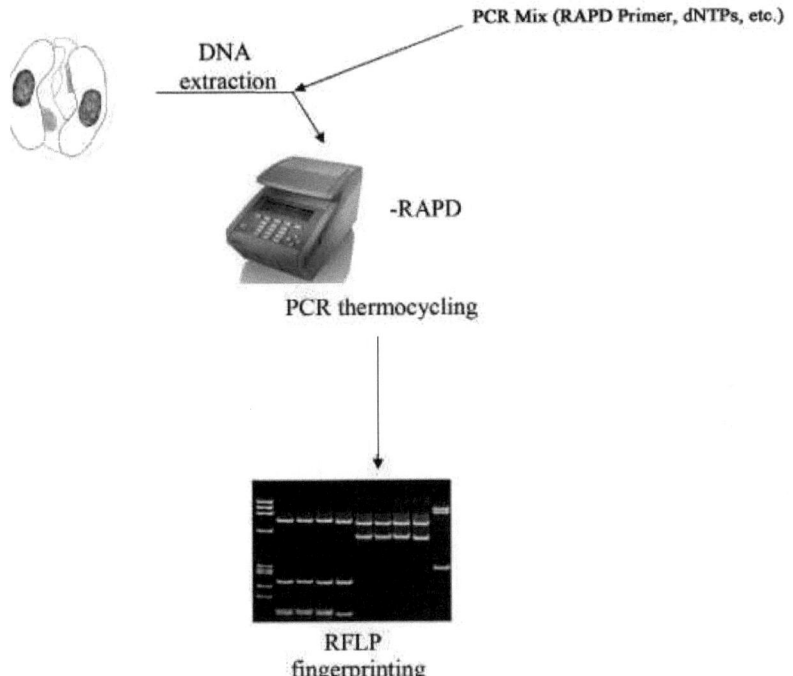

DNA extraction

PCR Mix (RAPD Primer, dNTPs, etc.)

-RAPD

PCR thermocycling

RFLP
fingerprinting

EXPERIMENT-15

Aim: Determination of G+C of Isolated DNA.

Theory: DNA is double helical structure consisting of two strands wound each other consesting of four nitrogenous bases Purines- Adenine (A) and Guanine (G), Pyrimidene- Cytosine(C) And thymine (T). According to base pairing rules, A always pairs with T by double hydrogen bonds and C pairs with G by triple bonds. DNA with greater G+C content will have more hydrogen bonds strands separate only at higher or greater temperature. i.e. it will have a higher melting temperature. When the curve is plotted, a plateau and the mid point of curve is the Tm and it is the direct measure of G+C content. G+C content after determination from the Tm of DNA. The G+C content calculated, thus, is very important taxonomically as is usefully less than 10% ever though the content may vary greatly between the glens

Requirements: water bath, thermometer, isolated DNA sample, test tube.

Procedure:

1. The isolated DNA is dissolved in saline, or 5ml of TE buffer use to dissolved and take into three test tubes labeled as 70°C, 80°C and 90°C.

2. Each test tube as placed in water bath at a respective temperature for 10 min. or 30 min.

3. Immediately take the OD at 260 nm.

4. A graph of temperature is taken at X-axes and then OD was taken or Y-axis from thus the Tm of DNA was calculated.

5. GC content is usually expressed as a percentage value, but sometimes as a ratio (called **G+C ratio** or **GC-ratio**). GC-content percentage is calculated as

$$\frac{G+C}{A+T+G+C} \times 100$$

Whereas the AT/GC ratio is calculated as

$$\frac{A+T}{G+C}.$$

Observation- The melting temperature and % G+C content was calculated with help of graph (Tm/OD)

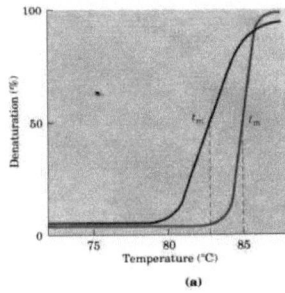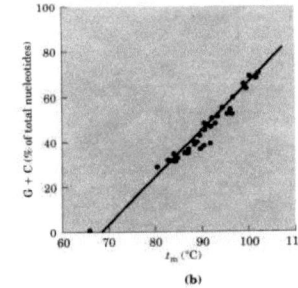

(a)   (b)

Precautions:

1. The labeling of test tube must be done properly & carefully kept in water bath of respective temperature only.

2. The OD should be measured immediately after the test tubes are removed from water bath.

3. Graph should be carefully plotted.